GOOD ENOUGH

A Truly Traumatic Tale

KERRY ROIS

ISBN: 978-1-6847-0664-8 (sc)
ISBN: 978-1-6847-0663-1 (e)

Library of Congress Control Number: 2019908335

Lulu Publishing Services rev. date: 06/21/2019

I would like to begin by stating that while this information is all true, it is a combination of things I was later told as well as things that I actually remember. This is being written (or at least started) one year, two months, and ten days after the day I nearly died.

I would also like to preface my autobiographical account by telling or perhaps reminding the reader that since no two brain injuries are alike, there is a very good chance that the issues I had to deal with are nowhere near what you or a loved one / friend had to work with. All I am able to do is provide an accurate description of the various struggles I personally had to put up with. Though you may be unable to relate directly to the issues I have presented in this text, it is my genuine hope that you as the reader can take a certain level of understanding from this work.

Contents

Introduction

For this series of events, I will start at the very beginning as I'm led to believe it's a very good place to start. The very beginning starts all the way back when I was an infant. Between my two siblings and me, I was not only the smallest but also the only child with life-threatening illnesses. Those ailments are none other than asthma and some serious allergies. These two disorders are fairly common, especially nowadays, when they are much more prevalent than they used to be. Thankfully, my mother is a nurse, and therefore, I knew what foods I was able to eat and what foods I was unable to eat. I also had an EpiPen starting at a very early age. Thankfully, I never had to use it until I was in my twenties, though it was always on hand if needed.

I never left the house without having the necessary medications on my person. I kept them in a fanny pack. I would take my fanny pack with me whenever I went anywhere for an extended period of time. In the fanny pack would have been my EpiPen, a new box of Benadryl, my inhaler, and a bottle of albuterol. My various family members knew about my food allergies and did their best to accommodate my needs. I also had an allergy to dust. And then, of course, there was my asthma.

Once I was old enough, I began to get allergy shots, and I began taking an inhaled steroid in an effort to combat my asthma. The inhaled steroid worked wonderfully on my asthma, and after a certain point, I didn't even need to use my inhaler anymore. I also had a nebulizer.

Though it was necessary multiple times when I was a child, I no longer needed it at a certain point.

I got the allergy shots once a week for several years. I don't remember exactly how old I was when I began getting them, but I do know I got them while I was a freshman in college. After I had gotten the shots enough times and I had gotten all the necessary injections, I was tested again with regard to the severity of my allergies.

When I was a child, I was so small and had to be tested for so many different allergens that the test had to be done on my back. Once I was in college, I was still small, but the test was able to be performed on my arm. For those of you who aren't familiar with this sort of test, they essentially put a small amount of an allergen on a needle and then insert it into your skin to see how your body reacts to the allergen. The more the area swells up, the more allergic to that allergen you are. In my case, most of the allergens had no effect on my body whatsoever when I was tested as an adult, which was drastically changed from how they would have swollen when I was a child. One important allergy remained, however, and that was my severe allergy to nuts. I was, and remain, very allergic to peanuts as well as all other tree nuts, including pecans and walnuts and foods like that. Two foods I'm not allergic to that are sometimes considered nuts are coconuts and almonds.

One of my other severe food allergies that thankfully went away was my allergy to seafood. I was so allergic to seafood that it wasn't even called an allergic reaction I would have were I exposed to it. It was instead classified as an anaphylactic reaction. Though this word may sound highly sophisticated and technical, it essentially means that my throat would swell shut in response to exposure to such an allergen.

I'm thankful that I grew out of this allergy because now, possibly as a direct result from not being able to eat it for so many years, I absolutely love seafood. I'm a huge fan of sushi in particular.

Also, once I discovered most of my allergies were gone, I was able to stop wearing the medic alert bracelet I had that alerted people to my allergies. I got it one year as a gift. I wore it every day for many years.

I haven't worn it for years, but I still remember exactly what it said: asthma, allergic nuts, anaphylaxis to fish. The bracelet charm didn't have enough room to list all my allergies, so my mother chose the most severe allergies to be featured. When I was a teenager, my older brother used a sharpie to add a comma between *allergic* and *nuts*. We both thought it was funny that because of the added alteration, the bracelet also said that I was mentally unstable.

Perhaps it's because this is just the person I was and therefore became used to, but I still don't think it's all that odd to have or have had such medical conditions. Neither of my siblings had such ailments, however, and that is slightly odd.

Years passed, and I went on with my life fairly successfully, not ever needing to use my EpiPen for any severe allergic reactions for the vast majority of my life. When it came time for me to go to a university, I initially selected one on Long Island. I liked it there and made plenty of friends, but after my sophomore year, I had to transfer schools. The main reason for this was money. Since it was a private university as opposed to a public one, it was simply more expensive than my parents were able to afford, despite my scholarships. I ended up transferring to a state school that was much more affordable as well as several hours closer to home. I was sad to be leaving the first college where I was accepted, but after I was at the new college for a decent amount of time, I ended up liking it there even more. What I learned after going through such a transition, is that it's much more about the friends you make and the people you know as opposed to the school itself.

In the summer in between my junior and senior years at my new school, something interesting happened. I was home for the summer, and some of my friends from high school planned on getting together simply to hang out. I was only twenty at the time, but we were drinking. One of the girls I knew had a friend who was the bouncer at a bar, and he let us inside. While there may have been many people I went to high school with there, there were also some I hadn't met before. They were the new college friends of one of my high school friends. Since I have always been a polite person, I did my best to try and make friends with these new people.

Once I was inside the first bar, one of these boys got my attention and offered me a drink. At the time, I was not savvy enough to know that was a horrible decision to make. I heard later that this boy most likely drugged the drink that he offered to me.

The next thing that I remember, I woke up inside of a cab with a random boy sitting in a seat in the row behind me. I had no idea what was going on, but I could tell by what I was able to see out the windows that the cab was on the highway and we were headed somewhere outside of where I lived. I guessed our location based on the signs, and I tried to call one of my roommates at my new university. It must have been too early though, because she didn't answer.

The driver pulled over to the parking lot at a rest stop, and his wife, who was with us, got out to use the restroom. I asked the driver if we could go back to where we came from. His response was that we certainly could. Once we were turned around, it took quite a while before we were back in the correct part of the state.

Once we were back in a more familiar area, the other passenger in the cab with me just had the driver drop him in a parking lot on the corner of two specific streets. Once he had gotten out and the cab pulled away, I began to cry. I was so confused and had no memory of how I'd gotten there. The cab driver's wife then asked to see my driver's license to determine where I lived. They took me back to my parents' house, and I hugged them both. My father gave the cab driver a generous tip on top of the money I had apparently already paid for the ride.

My parents both understandably had many questions for me, and my cell phone was inundated with missed calls and text messages. Once I was back in a familiar area, I was finally able to respond to the text messages and call back those who had called me and attempt to explain what happened. My mother asked me if I wanted to go to the hospital. I figured that I might as well, just in case anything bad happened to me or I had ingested a substance that was dangerous to my health.

The next day was a Monday, and I was told that some cops would be coming to the house to speak with me. Of course I understood, and I waited patiently for them to get there. Both of my parents were at work that day, so I was the only one there to greet them.

I remember that both of the cops were female, and we sat at the kitchen table while they asked me all of their questions. The only question they asked me that I can remember is whether I thought I'd been raped. Of course, I couldn't answer with any level of certainty, though I definitely didn't think so. When I told them that, I was told that I was lucky. They told me an anecdote about someone who wasn't so lucky.

This happened many years ago, though I honestly think it is an odd coincidence that once again, something negative happened to me that involved cars in some way.

The Day the Music Died

It was morning on Wednesday, January 14, 2015, and I was driving to work, so there were no other people in the car with me. Four months and six days after my twenty-fourth birthday, I was in a car accident that left me in a nearby hospital from January through Mother's Day. Apparently a semi-truck was pulled over onto the shoulder of route 33 / Kensington Expressway because the driver was somewhat lost. It was clear to everyone who examined the scene that he had been there for less than ten minutes. It's the law that after ten minutes, the driver needs to place some sort of reflectors to make other traffic aware of the truck's presence. As he had yet to do so, or possibly due to some uncontrollable environmental event, my car smashed into the side of the truck. My little red sedan was reduced to scrap metal.

Very shortly after the accident, I was taken to Erie County Medical Center, where I remained for several months. I received a three on the Glasgow Coma Scale, though I may have been already a bit sedated at that point. In short: the Glasgow Coma Scale is essentially a scale used to describe the level of consciousness in an individual who has sustained a blow to the head. 3 is the lowest score one can be given, while 15 is the highest. Not to be overly morbid, but anything under 3 is essentially dead or very close to death. I was in a coma for about two and a half weeks when I first was placed there. Once again, I don't know whether this coma was natural or a result of the sedatives I was given. For a while, I was completely optimistic about what might happen in the long run since the neurosurgeon had given me a prognosis of a full recovery after about two years. I naively assumed that my worst problems would be magically fixed once the two-year mark came and went.

I assume that my driver's license was on my person at the time of the accident. However, I was labeled as a twenty-year-old Jane Doe. Somehow the police were able to find where I lived. As a result, they placed several large, plastic bins on the porch of my home that were labeled incorrectly. I did not see these, though my mother saw them when she went there to collect some of my personal items.

Civilian Life

Exactly one week before my hospitalization, my parents had decided to get separated. They came to the hospital individually. A friend of mine told me they were no longer together, but I still had no memory of any of that happening. The day I was discharged, I had to ask my mother why my father didn't live there anymore. My memory, as it happens, has a rather large space missing. I remember nothing about the actual day of the accident, nor do I remember things that happened several weeks either before or after the accident. That my parents were apart made life slightly more uncomfortable and awkward for me. Essentially, I was in the middle of their arguments and disagreements on top of having to focus on my recovery. They would use me as a way of communicating between each other and do things with me or for me to try to make me side with one or the other of them if such a decision ever had to be made.

I recently heard the dissolution of a marriage described as traumatic for the children involved. At this time, my father was dating a woman he had dated before he and my mom got together. He also considered this woman's brother a close friend he'd originally had many years ago. He'd often mention him instead of his sister, which was slightly comical because I was well aware that his intention was to mention his newfound connection to that family as a whole. This might be why I first began to think about how selfishly my father had been behaving lately. I'm not saying at all that his attention should have been focused entirely upon me, though I had to deal with the aftereffects of the traumatic event that occurred in my life. Each and every day tends to be a struggle of some kind for me, so I would rather not be forced to hear about my father's girlfriend.

It is also for this reason, among others, that I most likely will never again be the exact person that I once was. Yes, my father continues to spend time with me on certain days, and on those days, he may take me somewhere. But I'm convinced that his absence will have a profound effect upon my personality. Recently my mom said that I was the most like him out of my siblings, but since I now have to essentially start everything over, that means I won't be at all like I was. I hear from friends and family members all the time that I'm the same person I was before the accident. While that is a nice sentiment, I don't believe it at all. My personality is now going to mimic, at least somewhat, my mother's rather than a combination of my mother's and my father's. It is also likely that hearing such comments repeatedly exaggerates the state of denial in which I find myself.

I've read that one of the most important aspects of an individual's recovery from a traumatic brain injury is love—that of family members and friends, I assume. Due to the dissolution of my parents' marriage combined with the fact my father constantly references his new significant other, this aspect of recovery is a tricky one for me. Many, though not all, of my family members live in a different part of the country. Other people my age also have jobs, whereas I do not. As a result, I am placed into the elderly category at most family gatherings. Because the elderly people in my family have nowhere to be during the day, they also take it upon themselves to reach out to me and show that they care by asking me things along the lines of, "How are you?" The problem with this type of question is not that my mood is so volatile that it needs checking up on. The simple truth of the matter is that I'm completely dissatisfied with what has become of my life. Essentially, though you may feel obligated to make sure that the brain-injured individual is not depressed, it is important not to smother her or him with too many comments and questions. As far as I'm concerned, based on the questions I was asked and the way I was spoken to and the fact others seemed to go out of their way for me, I felt I was being treated as if I was a physically as well as a mentally delicate individual. Being treated like this bothered me immensely because while I did have certain physical deficiencies, cognitively I was in a far better state then many others with or without brain injuries.

Today it has been exactly one year since I was discharged from the hospital. Other people may say that I've come such a long way and that I've seen so many improvements and the like, but I honestly don't know if I have a similar opinion. Ideally, I'd like to be further along in my recovery, but then again, I'd like for this never to have happened. That thought is obviously a waste of energy.

It's also frightening that one year later, I still have to live with my mom, I don't receive Social Security yet, my vision remains awful, the muscles on the left side of my mouth are both tight as well as somewhat weak and uncomfortable, and I still have a large hypertrophic scar on my neck that can be quite painful at times. I honestly never thought I would only have a little over twenty-four years in which to live. I mean, yes, I'm technically and biologically alive, but most aspects of my life are taken over by things I should and can do rather than things I choose to do for my own enjoyment.

The State of Denial

Once I had just one scar that was very small and nearly invisible unless I pointed it out. When I was in preschool, another student pushed me off of a large wooden slide. It was a fairly popular activity and therefore had a long line. I waited patiently in line until it was my turn to go on the slide. I don't think there was anyone in line behind me either. Just when I was about to sit down and ready myself to slide down, some girl walked up behind me and forcibly shoved me sideways just so she would not have to wait for me. I landed on a linoleum floor and split open my chin. I vaguely remember getting stitches, but it was a long time ago. What I remember is that it was picture day very shortly thereafter, and I had a piece of white gauze on my chin because of the stitches. My mom had told the teachers earlier to keep that girl in line a little bit because it must have been clear that she caused problems. Obviously they didn't, because they were genuinely busy or because they didn't think it was an appropriate concern to express about someone else's child. I hope that as a result of what I went through because of her selfish actions, she and / or her parents were reprimanded in some way.

I now have several new scars that keep me from forgetting certain things. I have a large hypertrophic scar on my throat where the trach tube was inserted. I have a much smaller hypertrophic scar on my abdomen where the feeding tube was inserted and removed, and though I cannot see them, I also have some scars on my head from the neurosurgery I received. The trach tube may have been surgically inserted, but it was

Kerry Rois

not removed that way. I have a dreamlike memory of being in a store with my dad, seeing the trach tube coming out of my neck in a mirror, and not being happy with it. In this dream like hallucination, I pulled the trach tube out myself. Unfortunately, though what I thought I was seeing at the time wasn't real, I did reach over and pull the trach tube out in real life.

On the day I was officially discharged from the hospital, the nurses showed my mother a few things she'd have to do to allow me to live comfortably in my home. I remember one specific thing that happened. I used the toilet, and the nurse told my mom to flush the toilet after I'd used it. I don't remember that being necessary while in the hospital, but on that day my mom took it upon herself to flush the toilet while I was still sitting on it, which got my behind all wet. I didn't need that to be done for me anyway, but the thought of her doing that still pops up in my thoughts occasionally and amuses me quite a bit.

She does seem to try too hard on occasion to make sure I do things in the correct order and do not injure myself. She also tends to act as my eyes on occasion, which is extremely helpful since, even with glasses, I am not able to see very well. She also seems to have a whole schedule laid out for me that includes seeing other people at least once each week, as well as making sure that I go out in public once a week. My mother seems to treat me like a disabled child. I decided this a while ago, but I'll say it again: I need to live on my own once again. Having to live here is now kind of a detriment to my overall well-being. Over six months ago, my mom and I went to one of my older cousin's weddings. The evening before the wedding, I showered at the hotel. My mom was close by in case any assistance was needed. She did end up shaving my legs for me because my dress was on the shorter side and my vision was especially poor at the time. She unintentionally cut my leg while attempting that, something I haven't done even with my poor vision.

I do still live with my mother, though we have since moved into a much more manageable home. It is now very possible for me to easily go into my own room if I want to be alone for some reason. In addition, the new home is all one level; I no longer need to struggle with going up or down stairs at all.

At this point, I do feel genuinely sorry for my mom. She has had to deal with the end of her marriage, the near loss of one of her children, and the loss of a house she loved. Essentially, she has quite a lot of things to deal with at this point—more than an individual should have to experience. Though at some point in the future I may feel like I'm ready and able to live independently of her, I relate to her struggles in a small way at least. I feel bad for her, and I genuinely appreciate all she does for me. Granted, it did take me some time to get used to living with her again, and she had to get used to living with me also.

A person whose role also involved taking care of someone with a traumatic brain injury said she had to bathe the person she took care of. In response to that statement, my mom agreed and said that she had to bathe me as well. In my mind, bathing someone else involves them sitting in a tub with some water in it, and the other person is then responsible for scrubbing different parts of the person who needs help with some type of soapy sponge. It is true that my mom had to be nearby to shave certain parts of me if necessary, such as my legs or armpits. My vision and coordination being what they were, I did not feel comfortable holding a razor blade close to myself at all. My mom does not nor has she ever had to bathe me. The thought of it still seems strange to me.

Since I had not lived at home with my parents for quite some time, I was not at all used to having my mother be there constantly. Possibly as a direct result of that, I stopped thinking of her as a human being with fears and desires. Once again, the way I perceived her was a result of me being overly self-centered. Lately I have tried to put myself in her shoes more, since she'd had to experience so many negative life events. When everything is combined, it's quite a mess. I honestly am not able to imagine what it must be like to have to put up with multiple negative events all at once.

Additionally, I hear repeatedly about how I'm so young and therefore have plenty of time to recover. I also hear repeatedly about how after my next birthday, I'll be too old to be covered by my father's medical insurance. These comments are completely opposed to each other. Either I am too young or I'm too old. At this point I'd rather not have a birthday at all. Last year my birthday was interesting, to say the least. There was a party that celebrated my birthday, though in my opinion, I was too old to have a party. I did see some members of my family that I hadn't seen in quite a while, which was pleasant. My birthday was used more as an excuse to celebrate the fact I was still alive than anything else. It was at my grandmother's house, which is to be expected since she has a swimming pool. She also invited some of her close friends; these are people I'd met before. I certainly wouldn't have considered the group lacking if they weren't there. Overall, the event was a touch awkward for me, the guest of honor.

Possibly as a result of the extreme level of denial in which I lived, I was very against taking many suggestions that were made to me. They were given in an attempt to help me, though I tended to dismiss such suggestions as being insulting and disrespectful of my age. It wasn't until after I had dealt with my injury for quite some time that I began doing my own research and thinking differently about such suggestions. I began to realize that it did not help in the least to be so defensive. Earlier in my recovery, I remained convinced that my life would at least return slightly to what it had been before the accident. It wasn't until quite a bit later that I became much more understanding and appreciative of all that was being done for me.

Results of the Damage

For an extensive period of time, my left eye remained shut. When it finally did open, I was only able to see things if they were on my right side. Things on the left were essentially nonexistent as far as I was concerned. I believe I may have third cranial nerve palsy. My left eye doesn't move up or down as of yet, and my left eyelid is currently drooping a tad. Also, the muscles on the left side of my face are extremely weak. Additionally, the inside of my mouth, particularly on the left side, felt very strange for a time. While I was in the hospital, the left side of my face was very weak, and I also had a strange tingling feeling in nearly that whole side of my face, particularly the muscles surrounding my mouth region. All I had to compare the sensation to was what I experienced if I ate something I was allergic to. I was also convinced for quite some time that the teeth on the left side of my mouth were made of plastic since I was not able to feel them at all.

Those assumptions were not true, as it happens. The muscles and nerves on the left side of my face were numb. I suppose the strange feeling was the beginning of their return. Due to the aftereffects of the trach as well as the significant muscle weakness in those muscles in my face, specifically those near my mouth, my voice served as an extreme embarrassment for me.

I find it interesting as well as confusing that not a single medical doctor said anything to me about some of the other issues I currently have due to the accident. However, I suppose the reason is because the most standard way of addressing such issues is not by a traditional medical professional. There's also the fact the doctors genuinely did not know how my body would heal itself as time went by.

My left shoulder and my left hip are not exactly where they should be, and I currently see a chiropractor, who's attempting to remedy that for me. I do not know yet how long it could be, but at least he's addressing the issues at all, whereas all the others recommended strengthening the muscles. That may be all well and good, but what is the point if the bones aren't in the correct places? I have also come to discover that my hormones are imbalanced and most likely have been for quite some time. I've learned that this can wreak havoc with a person's ability to fall and stay asleep. Extensive periods of insomnia are something I deal with quite often, unfortunately. Instead of being given a recommendation to see a naturopath or someone who deals mostly with balancing hormones, I was given a prescription for a sleeping pill. I took it one day, and it did nothing. I stopped using it completely. There were other issues that were related to hormone imbalances as well. It is currently my hope that I can address such issues by seeing a naturopath rather than ignore or get used to the issues.

I did have appointments with the chiropractor a few weeks in a row. However, it seemed to me, especially the last few times that I went there, that the chiropractor wasn't able to achieve any significant improvement. Therefore I eventually decided there wasn't much of a point in continuing to see a chiropractor at all.

Three days after I was officially discharged from the hospital, I began going to an adult day care type of facility. I received needed physical therapy there, and I was able to leave the house, which was necessary. While I was in physical therapy, my left leg was very ataxic and would shake fairly violently as I walked. The physical therapist encouraged me to put more weight on that leg when it shook; I also found that if I flexed the quad continually while walking, it wouldn't really shake. I tensed the muscle each time I walked for several months in a row. I no longer need to flex that muscle when I walk, which is somewhat of a relief. On occasion, however, one or both of my kneecaps will shake. I have learned from a renowned physical therapist this is likely because my nerves are hyperactive at this point. These nerves are essentially going above and beyond what I'm requiring them to do.

My left arm is still somewhat uncoordinated, which is interesting in a way since according to a new voice therapist I see now, my vocal chords are somewhat uncoordinated, which affects my speech in a very general sense. I remember that one of the occupational therapists I saw a while ago suggested brushing my hair with my left hand. In theory, this was a good idea; I do not think that I was completely ready at that point. I I find myself doing or trying to do this these days since I'm physically steadier. When that suggestion was first made, I remember trying to follow the suggestion. I was in the bathroom and had to blow my nose with a tissue. When I went to throw the tissue in the garbage, I stopped and figured I'd do it with my left hand rather than my right. Before I even opened the cabinet door behind which the garbage can stood, I fell sideways toward the right. Thankfully I didn't hit my head or severely injure myself in any way, though I was angry. I gave up trying to use my left hand after that. I wouldn't necessarily purposefully avoid using it, but I just tended to do most things with my dominant hand instead of struggling unnecessarily.

I have also recently learned that the extreme lack of coordination is due to what is referred to as cerebellar ataxia, and there are medications available to help. I'm currently awaiting the arrival of some supplements that I recently ordered. These pills are meant to calm the activity in the brain some, which should be beneficial for mood and help with sleep. It is also meant to alleviate a lack of coordination. I have no scientific evidence regarding its benefits as of yet, though I do have high hopes for what it might be able to do for me.

In addition to a lack of coordination, my left arm also tends to bend inward at the elbow. In a way this makes some sense since immediately following the accident, I had decorticate posturing. What this means is that as a direct result of the car crash, both of my arms bent inwards at the elbows towards my chest. I remembered this because the word features the word 'core', which is something found in the middle of something else. So my arms were bent in, towards my 'core'.

Changes

I suppose it's fairly common for individuals with brain injuries to have drastic personality changes. I'm not sure if I could say that I experienced that exactly, but I will comfortably say that I did become extremely emotional and sensitive with regard to how others behaved toward me. I also became rather suspicious about many things that were said to me. This may also be a result of my hormones being imbalanced. Also, a traumatic brain injury may cause a lot of deficits and challenges, but it often has no effect on IQ. Apparently when you have one, there can be physical issues only or mental issues only, or you can have some of both. Based on the care I received directly after the accident, I was mentally all right, but not so much physically. Therefore many people, family members included, spoke to me and acted toward me as if I was either a young child or an elderly individual. This is also where my sensitivity and suspicion stemmed from.

Very shortly after I had neurosurgery, I was given a complete recovery prognosis. As the neurosurgeon said, this could happen in as little as two years. I was very naïve as well as gullible. Call it wishful thinking, but I believed for quite some time that after a relatively short timeout from what my life had ultimately become at the time, I'd be able to return to what I had built for myself. I only decided this would be a short timeout. For quite a while at the beginning, I considered it to be too long of a timeout from any sort of meaningful life.

While I have since learned that such a prognosis is not at all common for individuals with traumatic brain injuries to receive, I do find it somewhat intriguing that many of my difficulties were physical and therefore associated with my cerebellum rather than my cerebrum. And yet, despite the fact the damage may have been fairly obvious to the neurosurgeon, not a thing was said to me or my mother about the results that might be expected due to such abnormalities.

Young Adult?

My parents still lived in the house we moved into as a family when I was ten, but I hadn't lived there for several years. With my dad not living there anymore. It was just my mom and me. My dad was still nearby, and he would visit me two or three times a week. I received physical and occupational therapy five days a week from May to September. My main issue with the facility was the other patients who received treatment. In general, they tended to be elderly—so much so that the employees would often go through the patients who hadn't been there for some time and make a note if they had passed away recently or if there was some other explanation. The employees also seemed to go out of their way to find another patient who was at least closer to my age for me to interact with. She was only several years older than me, as opposed to many others whose age was north of fifty. However, her issues couldn't exactly be compared to mine, since she was born with her disabilities and had gotten used to compensating for them whereas the same could absolutely not be said about me. I also met with the social worker there a few times as she attempted to counsel me. I have since learned that it is part of every form of medical professional's education to learn at least a little about counseling people. I was also put in an area that seemed to be designated for individuals in wheelchairs because at that time I was also in a wheelchair. The difference in the issues that I had to deal with versus what they had to deal with were so extensive that I considered it a waste of time. In short, I did not particularly enjoy going to that facility, especially as often as I did.

Apparently after the first day I spent there, I said to my mom something along the lines of, "Don't make me go back there." Unfortunately for me, that was not a viable option at that time, and therefore I continued to go there for quite a while. In fact, it was kept up for too long in my opinion.

I'd been receiving physical, occupational, and speech therapy at least a few days each week since I was officially discharged from being an inpatient. After the previously mentioned facility, I received outpatient therapy at the same hospital where I was kept alive immediately after the accident. Currently it has been one year, and seven days shy of three months since the car accident, and I have been officially released from the facility where I was receiving therapy most recently. I will choose another type of therapy I'm sure. The therapists I saw every week for six months seem pretty confident that I will continue to improve without their guidance. I, on the other hand, think that I would benefit from more or a different type of therapy. Of course, I will continue to do beneficial exercises and the like, though I'm fairly sure that a new, different therapist might be useful as far as my recovery is concerned.

At this point, I'm somewhat concerned that my speech still sounds strange, though it's been over a year. My mom seems to think that there's an issue with my trigeminal nerve. This is because the muscles on one side of my face are mostly stoic. I am unable to produce tears at all, and I have continued issues with my vision in general, as well as the fact my left pupil is permanently dilated and the eyelid on that side is a touch droopy when compared with the right. In addition to the fact I'm unable to speak very clearly, my voice is also much more weak and raspy sounding than it used to be. This may be associated with the fact my vocal chords present different thicknesses. The difference was most likely caused by the accident. As far as I can tell, though others may disagree with me, my voice hasn't changed much since I became aware of things while in the hospital. For a time, I had some concerns about how my sessions with the speech pathologists unfolded. I do realize that speech pathologists work with memory quite a bit as part of the work they do. However, since I was given a complete recovery prognosis by my neurosurgeon, I would think that they'd have made more progress if they'd focused more on my actual speech. They may have touched on that aspect slightly. I do think it's very possible, though, that they were able to recognize that my strange speech was related in some way to nerve damage, and if that were the case, little could be done on their part.

My voice continues to cause me realistic problems in terms of how old people think I am. My voice is an issue, combined with the fact I'm very small in stature and cannot drive and am therefore nearly always accompanied by one of my parents. Because others assume that I'm quite a bit younger than I really am, people tend to speak to me in a very condescending or patronizing manner. Unbeknownst to them based on my physical appearance or voice alone, I am in reality a young adult who is somewhat more intelligent than others my same age, possibly even more so than the speaker.

Living with my mom was great at first, especially because I couldn't see or do many physical things that well. I do have certain food allergies. Because of my lack of decent vision, I ate a bowl of my mom's cereal that contained something I was allergic to. I called my mom, who came to pick me up and take me to a nearby urgent care facility. Whereas I would have required a shot of epinephrine in the past, I was now able to calm it down slightly with a few Benadryl. Though it was not as life threatening as it once had been, my food allergy still existed, but are insomnia and poor short-term memory a fair trade? My mom also tended to frustrate me with things she would do or say or the ways in which she'd behave in general. I desperately wanted to live independently again after a time, though I did have to wait to be approved for the traumatic brain injury waiver before that was an actual option for me. I also learned recently that even with the waiver, I would not be disabled enough for the traumatic brain injury subsidized housing to be an option. I am also convinced, however, that living on my own would allow me to recover at a somewhat faster rate. I'm well aware that my mom handles many things for me, some of which I think I should handle on my own. Once while driving me to an appointment, she got lost and had to pull over in a parking lot to figure out how to get there. I ended up being late for the appointment and had to wait a while longer before the doctor was able to see me.

I also recently had an appointment with a gynecologist. My mother had an appointment at the same facility at about the same time. After the appointments were finished, she told me that she actually considered going into the exam room with me. Thank God she decided against it, because that would have been extremely uncomfortable for me. She tends to do that sort of thing fairly often, however. She'll attempt to help a situation in some way by providing physical aid or by placing herself in a situation where I'm asked a series of questions. Very often, I think it's overkill. Such interventions, while they are meant to be helpful, end up leaving me feeling uncomfortable and embarrassed.

Kerry Rois

Version 2.0

I suffer from ataxia or shaking in my left arm and fingers as well as my left leg. I was once a singer in an acapella group in college. I also took dance classes from age three every year up until I was eighteen. At this point, attempting either would be a comical undertaking. My balance is lacking, and I can't even speak that well. However, I suppose it is an improvement somewhat seeing as how when I first started speaking after the accident, I would only speak in French. In a way, I suppose it was my brain's way of trying to hold on to certain things. After all, I did take French for six years. I also considered switching my major to French since I did receive a college credit for taking French 6 AP. I had a meeting with the head of the foreign language department at the school I went to at the time. He only spoke in French, and I was meant to do the same. It didn't go very well since despite all the years of French I'd taken, I was not fluent in the least. I later selected a different major that I enjoyed, and I was able to graduate with a magna cum laude distinction.

I also go through bouts of extreme depression and suicidal thoughts. In my opinion, things would be much simpler if I had passed away as a result of the car accident. And yet, I have no other option but to act as if I'm in a pleasant mood at all times.

I tend to dislike certain generalities that are made regarding traumatic brain injuries. All brain injuries, be they traumatic or acquired, are truly unique, after all. In my case, I had a few family members who took it upon themselves to learn about those symptoms that were fairly common among traumatically brain injured individuals. I would sometimes be asked questions based on the assumption that such symptoms were true for me. They tended to be untruthful in my case, and I would be insulted immediately following these questions. One result that I've heard is common among brain injury victims is that having a very strict, highly scheduled day is helpful. As far as I'm concerned, this is absolutely true.

For quite a while, especially around one year after the accident, I became very suicidal. At the same time, I was in a great amount of denial. As previously stated, I felt very strongly that it would be so much simpler if I was dead. I tended to speak to my immediate family members about these suicidal thoughts. However, they would become extremely worried and upset by the suicidal thoughts. Occasionally the thoughts would be significantly more severe than they were at other times, though they did plague me for quite a while. I also was not ready in the least to accept the fact I would most likely be a slightly different person than I was and I might not be able to do all of the same activities I was once able to do. In my head, I seemed to revert back to when I was a senior in college. It was a genuinely positive time period in my life, and so it makes sense that I would mentally revert back to that specific era.

Additionally, so many people were giving me things I didn't need and taking me places. I often felt guilty about all of the things I was being given. I also felt like even more of a disabled person because of all these things. In my case, the sheer multitude of gifts made things more stressful in a way. I already had the things I needed to live since I had been living on my own for several years before the brain injury. Additionally, it was me and my mom alone in a house that was originally purchased for five people to live in. There were also a number of things that remained in the house that were the property of people who did not live there any longer, some of whom no longer lived in the same state. I could handle being taken places, though I was not very comfortable with myself once there.

I was also spoken to or spoken of within earshot as if I was a seven-year-old or an eighty-seven-year-old, rather than a twenty-four or twenty-five-year-old, which is what I actually was. I was often spoken to in such a way by one of my own relatives. Several months shy of a year, this relative visited me at least once each week. Often she'd bring along her best friend, and once she brought the man she lives with these days. In my opinion, she relished the fact I was constantly at home, and therefore it was fairly simple to force visitation upon me. Her best friend I could certainly stand. She's very nice as well as understanding. The man who lives with her these days is another story. He is not related to me in any biological way, and some of the things he says bother me to no end. He even asked me for a kiss once.

Thinking back on the event, I could have made some kind of noise while keeping my mouth closed and said, "Sorry, I just threw up in my mouth a little."

That is how I felt shortly after the event. I do that often these days: I think of a clever comeback a while after a reply is no longer necessary. I have a strategy that could work in my favor as well as keep those individuals who repetitively insult me fairly calm. I will simply act apathetic in most ways and amenable when appropriate. Short questions that only need one-word responses, I'll continue to answer. All other questions and comments I'll ignore or answer with nothing more than a shrug. I suppose this is how I might act toward or around people at times when I feel particularly negative, like when I feel like the other person most likely doesn't care much or at all about how I actually am feeling or doing.

Age Does Matter

Being spoken to as if I was an elderly individual was interesting, since that was an accurate description of many of the individuals I found myself encountering on a regular basis. Since I have a neurological disorder to deal with, I'm often placed in the same categories as elderly people who've had a stroke or something similar. I'm honestly worried at this point that I'll quickly become so accustomed to interacting with persons of that age group rather than persons of my own age group and that my personality will be affected in a rather strange way.

I became rather suspicious of most things that were said to me or about me. Even if they were a fairly common thing to say, I assumed that they were allowing my current disabilities to influence their actions or comments toward me. All too often, I'm spoken about well within earshot, as if I'm unable to speak for myself. I'm certainly able to speak for myself. Sometimes I simply do not willingly offer up such information, because I don't think it's necessary that I do so. I also became suspicious of people close to my own age group—call them friends or peers or whatever you'd like. Some of those acquaintances of whom I didn't think much decided that I was a completely different person with a completely different group of opinions. I suppose this could also be the direct result of the fact that, despite having no real thoughts about them, I was able to act as if they were very important to me all along.

I have also recently discovered that elderly people seem to be worse than others as far as understanding my condition is concerned. I assume they often recognize physical issues that I am currently experiencing and relate it to themselves or someone they know who is of a similar age to themselves. As a result, they will often make suggestions directly to me that I think would prove less than effective. In my opinion, these suggestions would serve only to treat specific symptoms I'm currently suffering from rather than the actual problem. I generally tend to thank the individuals who make such suggestions and reply that I'll look into it. They obviously don't really care what I think anyway, or at least that's what I assume.

Free Time

Once whenever I had a moment of free time, I was able to do something out of the ordinary that I actually wanted to do. Currently I have too much free time. Unfortunately for me, I do not spend said time involved in things I want to do; instead I find myself engaged in activities that I feel obligated to spend my time practicing. I do not know if or when it might be, but I would feel amazingly relieved should a time come in which I could actually plan days where I could do activities that I enjoy.

People often tell me things like, "You're so positive. It's wonderful," and the like.

Whenever people ask me how I'm doing, I generally respond with a quick, "I'm fine, and you?" or something very similar.

In my opinion, people don't really want to know details or specifics about why I'm depressed or embarrassed or why I hate my life. It's a big waste of time to ever make those kinds of statements whenever I am posed such a generic question.

Recently I've thought about what an unlucky life I've lived. It began with splitting my chin open when I was three. I was also drugged by someone at a party and awoke in a cab over an hour from where I lived at the time. I was also taken to the hospital once after fainting on the sidewalk a few months after graduating from college. It has continued with this traumatic brain injury fiasco, which was someone else's fault and not my own. I may never understand how a single individual can be forced to withstand multiple traumas.

Kerry Rois

When I was in college, I remember playing some sort of drinking game (Never have I ever, perhaps) in which you would say something that was untrue about yourself. If that statement were true for anyone, then they would take a drink of their alcoholic beverage. If you were up, you'd say something that was untrue for yourself to make others drink. My friends would select this because none of them had any scars to speak of and could therefore talk about scars so others would have to drink. It was during such times that I would briefly explain where my scar was and how I came to have it. Now if such a game were to be played, it would be quite obvious to everyone involved that I had at least one scar. Many of my scars I cannot point out to others because while I know they're on my skull, I don't know their specific locations. The two I do have knowledge of are hypertrophic and are painfully obvious. They exist because of my breathing tube, as well as my feeding tube. The large scar on my neck is more obvious and also more sensitive. I've often covered it with Band-Aids in the past, although because of my lack of coordination, particularly in my left hand, I only wear a scarf now if I feel it should be hidden.

Posttrauma

I'm well aware of the fact my postinjury situation could have been worse, though that does not make my circumstances easier to deal with. I often contemplate suicide and have even attempted it on multiple occasions. The attempts proved to be futile. At times when I feel emotionally more stable, I'm glad that I was not successful. There are certain instances, however, during which I am convinced that I'd be better off had I not lived through the car accident. My family members would not be better off in the long run, but fewer tasks would have to be done simply for my sake, and more money would be saved as a result.

I also recently read about how being surrounded by one's peers can speed up recovery some and is less likely to result in depression. This I can understand, though I also feel as though I've basically been sabotaging myself this whole time so far. I've been spending much time with my mother as well as other family members by association. While I can't say that some of these family members have had any sort of negative impact, I genuinely believe that some have. It's also difficult to have people drop by very often because they have careers and things to do. I, on the other hand, do not have a career or any source of income at this time. Some of you reading this may think after making such a comment, *What about Social Security?* The reality is that I've been denied even that type of income. Either we go somewhere and they foot the entirety of the bill, or they come spend time with me at my current dwelling. Each of these options is less than ideal as far as I'm concerned.

Kerry Rois

Weekends seem to be the most difficult for me. I honestly do not know if this is because my mother is home all day or if it's because I am reminded more of the life I no longer have. Weekends had once been a time of rest and relaxation, a time when I could do the things I actually wanted to. My life currently is a different story. I have no income, and therefore I cannot comfortably suggest a location that will require some type of expense. Also at this time, I'm not very comfortable with myself and therefore not very comfortable being around others. I can't say that I've ever been an arrogant person, but I was fairly confident as well as comfortable with who I was. And though I'd become comfortable with that person (me), I had been forced to get used to an entirely new person altogether. I wasn't a completely different person, but I was slightly different. I was not at all comfortable with the fact I had to become used to the way things were.

I was also in a state of denial for quite some time. I was convinced for some reason that eventually I'd be back to my old self, and therefore I refused to learn to do things in a different way, thinking I'd be back to myself in no time at all. It seemed that these thoughts were most likely encouraged by the prognosis I received from the neurosurgeon. He probably had a good reason for giving such a prognosis, but my current thought is that he gathered it from the quality of my brain, my cerebrum in particular since that is what he focuses his medical attention on. As far as he's concerned, the physical ailments could be remedied either with physical therapy of some kind or by the use of some sort of assistive device. I suppose this is true, though I never imagined that my life would end up as such, and I also never imagined that I would be the victim of such a catastrophic incident.

In four days, I will be twenty-five years and eight months old. Never would I have thought that by this age I'd have so many regrets, similar to those that elderly people nearing the end of their life experience, and yet I do. I frequently think of things I might have done differently to avoid this event from ever occurring. And yet, people expect me to be in a pleasant mood. In fact, they often make comments about my demeanor.

Little do they know that in reality I'm quite miserable and I hate what I've become and I wish that the accident had taken my life. But for all intents and purposes, I must act as though I'm in a wonderful mood. I'd rather not have to go on like this at all, and yet I must. Not for my own sake, of course, but for the sake of my family members. As a result of how poorly I feel, I'm genuinely afraid of myself and what I might do. After all, this is not an existence I'm particularly fond of.

As previously mentioned, I have no job, and I do not drive. Some days I find that I have a purely adequate level of motivation; on such days, I have my own schedule made out for myself that touches on different types of exercises and other activities I know I should do. On other days, however, I have a complete lack of motivation. The boredom on such days has gotten the best of me several times. Very possibly as a direct result of such boredom, I have unsuccessfully attempted suicide on multiple occasions. Immediately following such failed attempts, however, I regret the fact I tried at all. I'm also not a fan of blaming things on the fact I'm brain injured, as that is almost too simple as far as explanations are concerned. In this case, I do think it's an accurate explanation. At this point, since I've been living this way for over a year and a half, I do not often consider suicide. Yet I also feel as though I'm a burden to my family members. I honestly thought that having me completely out of the picture would make things somewhat simpler for them. In general, I'm not exactly pleased with how my life is now, though I rarely take matters to such an extreme level.

I am now twenty-six and would also like to say that while I may have been unrealistic for some time about my recovery, I wasn't very happy or even contented with the way my life was going immediately leading up to the accident. I had two jobs, though I strongly disliked the one that paid me more money. I wasn't all that fond of the fact I still lived in the region in which I was born and grew up. I drove a very old, broken car. Overall, I didn't have much of anything to really motivate me to get back to living the life I had cultivated for myself.

I made a remark earlier today that was meant to be painfully true while also being ridiculous. I graduated from college with a magna cum laude distinction, and my mom graduated from college with a summa cum laude distinction. The comment I made was that because we received such honors and yet were dealt such less-than-ideal lives, maybe we're too smart for our own good.

Attempted Sequel

Unfortunately my family members are not willing to cope with the changes I'm undergoing as I become more independent. I fully understand that people my age or close to my age have jobs and their own lives and things, and yet there's the living with my mother ordeal. She seems either unwilling or unable to take a step back and allow me to deal with things on my own. There is also the notion that most, if not all, the other adults treat me or speak to me as if I'm a child. And as if I didn't have enough to deal with, there's the fact my parents are not together anymore and treat me as if I'm something they need to fight over.

My father visits on specific days. He rarely suggests that we spend time together on a different day of the week. Additionally, once when I was out somewhere with him, he had me meet his current girlfriend. He still claims that it was merely a coincidence, but I'm not entirely sure that I believe him. Once again, my disabled state was being taken advantage of. Now that I think about it more extensively, he has suggested that we spend time together on a day that was off his routine schedule. He sometimes will also not visit on days when I expect him to. This frustrates me some, especially since at the time of day when he generally drops by, I purposefully choose not to do anything that is a solitary activity. Similarly, I was recently told over the phone that I could expect a transportation service in "a while," those were the woman's words. It may have taken a few moments, but eventually I became angry. I'm very easily angered when specific details and explanations are not given and the situations or relevant time frame are generalized. Whether this is a result of my brain injury or not I cannot say; I'm sure that some, including my mother, will say that it is.

As far as I'm concerned, I would rather not have a remarkable brain, especially while at the same time having numerous physical issues. I'm able to remember all of the things I was able to do in the past. Even thinking about such things and comparing them with the struggles I currently have adds a bit to my depression. Additionally, this causes people to speak to me and treat me in a manner that is solely reflective of what they are able to see. This upsets me greatly since that my physical abilities are drastically different than my cognitive abilities. It often results in people speaking to me or treating me in a way that is intensely untruthful as far as the actual degree of my injury is concerned. In fact, I took an IQ test very recently that resulted in me getting a score of 139. In an attempt to learn more about what that number meant, I looked up what different scores signify; apparently 140 and above is genius level. Therefore, I'm a brain-injured genius.

My mom also fairly often refers to me as being selfish. After thinking about it more closely, I have come to the conclusion that everyone is selfish some of the time. My instance is slightly different than other people's. Constantly I'm spoken down to or given too much help. I will admit that on occasion, I do require a certain level of aid. It is of course true that I do not know what others may be thinking at any given time. And yet, based solely on how these others speak to me, I assume that they are placing me into a category of individuals who require an endless amount of encouragement and positive feedback. This all tends to bother me quite a bit, especially since I'm perfectly able to see that such comments are completely void of reality. Since the way others speak to me tends to bother me, what I'd prefer is for others to generally leave me alone.

I have been told that I tend to isolate myself quite a bit; I fully agree with that statement and am attempting somewhat to remedy that. It is pretty difficult, however. People tend to treat me in a way based on the issues that are visible to them, which I find insulting. When multiple individuals are gathered in the same area, an issue that has also arisen is rather than me isolating myself, the other individuals tend to isolate me. This is understandable in a way; their desire is to move around freely and take part in various forms of activities, whereas it takes me much more effort to move around or even to walk at all.

Kerry Rois

Troubles

Speaking on the phone tends to be problematic for me. At one point, it did prove rather difficult for me to understand the individual I was speaking to. If this was a relative or someone who was familiar with the sound of my voice, it wasn't that problematic at all. If, on the other hand, I was making a phone call to someone who was unfamiliar with my voice or accident, I was often transferred to someone else. If a telemarketer called and I decided to actually answer, they would often hang up on me. In my opinion, this business with the telemarketers proved to be beneficial for me. What tended to happen with regard to calls I made discouraged me quite a bit. The reason I made certain phone calls was an attempt to gain a small amount of my independence back. After making several attempts at calling a specific number, I would often give up and ask my mother to take care of the matter on my behalf.

My emotions tended to get a bit out of hand, mostly with regard to being depressed. I still found certain things enjoyable; however, I also tended to feel like I had become a ridiculous person who should probably have died in the accident. Occasionally my suicidal thoughts would frighten me, because I honestly didn't know if I would attempt to carry out any of the thoughts I was having. My mother has always been very against medications; therefore I was on none in the beginning, and that could have inflated my emotional landscape. I was at least slightly depressed before the accident occurred. Whether this was made more severe as a result of the injury, I do not know.

I have learned that such changes are possible. One biological change that I experienced is that while my allergies still plagued me, my singular food allergy seemed to become less severe than it had once been. Additionally, I used to sweat quite a bit to the point that I would have to purchase a specific type of deodorant for myself. I would generally put that on at night and use a different sort in the morning. After the accident, I no longer suffered from hyperhidrosis and could therefore use regular deodorant. Whether these changes are a result of the injury or the brain surgery, I do not know.

My emotional valleys did not seem to last very long. I would often feel extremely discouraged one day yet feel all right by the very next day. On days when I felt more at peace, I tended to think how silly it was to ever have been so suicidal. And yet, I was occasionally.

Kerry Rois

What Now?

I got mixed reviews upon my recovery and how it had progressed thus far. Some people I interacted with had very positive opinions on how I might recover. Their viewpoints were similar to what I believed would eventually happen. Other people had much more realistic and therefore more pessimistic viewpoints about what might happen to me in the long run.

I read recently that individuals with higher-than-average expectations for themselves tended to have a vastly more difficult time accepting the new person they'd become, especially when compared to the person they used to be and the specific tasks they were capable of. This is a category in which I think I belong. As a result, I'm genuinely unsure as to whether I'll have the ability to accept the new version of myself at all. Additionally, I read that highly educated people feel similarly about such a change. They feel as though all of the work they did to get them to the end point was a waste of time and was for naught. It is true that I have nothing beyond a four-year college degree. However, I did put a lot effort into my education in general. I received a magna cum laude distinction on my diploma, which is something I'm still quite proud of. In addition to being discouraged by the fact my brain had changed at least somewhat, I was also frustrated by how people would speak to me or treat me considering my former achievements. It seemed to me that most if not all other individuals assumed I was unintelligent as a result of the brain injury. This also seemed to strongly influence the sorts of options I was able to choose from. For instance, many jobs I could choose from while being a brain-injured individual involved secretarial duties of some kind. I have nothing against individuals who have made their careers in such a field; however, I assumed that graduating from college would offer me a different group of options. Perhaps as a result, I felt as though everything I learned in school was for naught. Essentially, I was convinced that I'd lost most if not all of what I'd learned.

I learned also that highly educated people as well as smart people in general are more likely to recover successfully and fairly speedily when compared to others who are more average in terms of intelligence. For this, I think I have my father to thank. He's not highly educated, although he is quite intelligent. A few years ago, he also got into the Mensa association. He mostly tried to break into it because his brother recently had become a member. I do not know what my IQ was before the accident, but I took an IQ test a few weeks ago and my IQ was somewhere around 139. Yet people still speak to me as if I'm stupid because of what happened. Since I learned that number, I have considered putting it on a sticker and wearing it when I'll have to interact with certain people who continue to speak to me like I'm unintelligent. I later thought that if I were to wear that, it would result in the other individuals stopping me, reading it, and commenting on it. I certainly wouldn't want to suffer through that, and so I thought better of it. It's also interesting that I got such a high score from an IQ test after it had not been two full years. Apparently that's the generally agreed-upon time frame in which the brain is able to make significant improvements.

One of the books I was reading that my sister had gotten for me said that a lack of control, or a sense of it at the very least, was apt to make people more depressed. It also said that a person requires autonomy or an illusion of it. Otherwise they are more likely to fail at a specific task and to become depressed to some extent. Unfortunately, as a direct result of my injury, I had very little control or autonomy over any aspect of what my life had transformed into. Such unfortunate realities were now very realistic for me in my post–brain injury life. I also recently read that people with brain injuries often feel as if they're burdens to others. They tend to become more self-centered than they used to be. I may have read other things that accurately describe people following a brain injury, but I remember the examples I have stated because I feel like they are certainly true for me. It is rather difficult to not feel like a burden when people buy you many things because they are trying to make sure that you are able to feel some joy following such a catastrophic event.

Remainder

I had a thought today that caused me to weep, or would have had I the biological ability to form tears. Apparently, many therapists and the like are finished with me since I meet all the requirements necessary to live independently. That level of independence is their goal for any individual. This is true, and I can see that. What truly saddens me is that my current life with all my capabilities and shortcomings, though they may change slightly over time, will essentially define my existence for the years that remain. That fact, while it may only be as true as I make it, is one I'm not at all fond of. The remainder of my life will most likely be somewhat of a struggle. It's one that I'd rather avoid, yet I must attempt to work with it. Apart from said struggles, I most likely will continue to be defined by this medical fact for the rest of my life.

I also tend to use certain things that are common for individuals with brain injuries to my advantage. For example, if a certain someone I'm not particularly fond of insists on paying me a visit, there is a strong chance that I'll pretend to fall asleep so they might be prompted to leave slightly sooner. I will often do things intentionally wrong and revert to the excuse, "I'm brain injured." This is of course true, and people often feel as if they cannot challenge me. I'm honestly not sure if this usage can be classified as good or bad. It might be good in the sense that I'm freely volunteering certain truths about what it means to be brain injured. It might also be inadvisable since I'm unnecessarily exploiting certain aspects of being brain injured.

Kerry Rois

Currently, though, it's been well over a year. I seem to be unable to stop comparing who I am and what I'm capable of now to who I was and the things I was capable of before my entire life changed for the worse.

I realize that many people will say things along the lines of, "It hasn't changed for the worse; it's just changed." That may be a nice way to look at what I'm currently struggling with, though in my opinion it is a change that has downgraded my capabilities somewhat. Similarly, I seem to have a lot of difficulty with being positive in any way, shape, or form. Yet I must keep going. As previously stated, I often have suicidal thoughts. However, such thoughts are merely another way I'm being selfish.

Almost a Year and a Half Later

I need to become accustomed to the fact I often forget where I have placed a certain object. This is particularly troublesome when I later decide that I need to use the item, whatever it may be. Additionally, I tend to forget the steps necessary to achieve a certain outcome unless they are clearly defined for me. As a result, I ask that the steps involved with a specific task are written down for my benefit. For this reason, following a recipe while cooking is not such a difficult task for me. I am nervous about the idea of having to hold a sharp knife or coming very close to a stove. Because of my physical as well as cognitive difficulties, I am not able to trust myself with certain tasks.

I don't think this will be a hindrance with regard to me needing to find employment of some kind; however, these things are a struggle that I will have to become accustomed to as far as my everyday life is concerned.

Since I am not currently employed and must constantly spend time with the elderly, either related to me or not, my life has become somewhat of a misery. I'm dissatisfied most of the time and by myself also. I cannot drive a car, and neither can I walk long distances without one of my hips hurting. As far as I'm concerned, though it may be different than what medical professionals believe, the car accident essentially did end my life. All the time I hear things along the lines of, "You get to start over. Life is what you make it," and things of that nature. I don't believe any of it. I have too many physical disabilities as a direct result of the accident to grant any such statements even a shred of truth.

July the Following Year

I may not be completely at peace or content with my life at this point, though I know it could be much worse. I suffered a traumatic brain injury, true, and that could not at all be considered lucky. It really did place a major shift in what I had built for myself as far as my life was concerned, and for what I was able to realistically envision for the future. That being said, I do feel somewhat lucky in a sense with regard to my recovery. Some of my family members have been unenjoyable to be around and inconsiderate to my current state. There are others, however, who have been amazingly helpful. These helpful family members have great connections with regard to my recovery. As a direct result, I've had the opportunity to engage in programs that otherwise wouldn't have been possible. The fact I am able to participate in such activities or use such devices whereas many other brain injury victims cannot does make me feel lucky in a way.

I feel as though it's my duty to reach out to other traumatically brain-injured individuals to offer them the same types of miraculous opportunities that I was offered. At the same time, however, I also tend to go about each day with a complete lack of enjoyment for life. I do occasionally slip back into thinking that I'd be better off had the car accident actually ended my life in a biological sense. I understand to some extent the importance of being out in public, though doing so tends to make me angry since I'm assaulted with views of others that are representative of how I used to be but no longer am. I have also been spoken to / questioned by more than one stranger. I would be asked if I needed help with anything, or if I was ok. I cannot say for sure why, though I do believe I was spoken to because of how I walk currently.

Also, though it has been almost a year and a half since the accident, I am not currently on antidepressants. This is because my mother was very much against it at first. At this time, other people whose opinions she trusts told her it would be a good idea for me to be on them. I also told her that my failed suicide attempt should make it necessary for me to be on antidepressants. I feel as though I should have been on them for quite a while at this point. My mother is still attempting to find an MD who is qualified and able to prescribe such medications. And so I say to you: brain-injured individuals may need to be on antidepressants. It's not a fact to be ashamed of. It is often necessary to have any sort of return to an average life. The mental turmoil that accompanies a brain injury, traumatic or acquired, is not something that should be taken lightly. Seeing a psychiatrist or psychologist on a fairly regular basis as well as taking antidepressants are things that should be given serious thought. Additionally, brain-injured people may need to fight for such methods of recovery to some extent. If your caretaker is against such methods of recovery, like mine was, you are likely to have a much more miserable existence that is drawn out. I can't at all speak to how taking antidepressants can change things. All I know is that at this point I do not take them, and I am miserable with each day that passes. Though this may be completely inaccurate, I have an idea in my head that once I am able to begin taking them, things will become infinitely better and more positive.

That being said, I no longer wish to seek out antidepressants. One of my cousins gave me an oil diffuser and a small vial of frankincense, which I've been trying to use almost every day. There's also the fact the supplements that I'll be receiving by mail soon may improve my mood. Overall, I no longer think that seeking out antidepressants would be a productive use of my time. I've also taken it upon myself to do a little research on the frankincense oil and found that it essentially is a natural antidepressant. The types of pills that I received through the mail and have now been taking for a few days are a neurotransmitter called GABA. I have learned through my own research since ordering them that in pill form, they cannot pass through the blood brain barrier. It also works best to take them on an empty stomach, about two hours after eating. Since I've been taking this type of supplement, I have been feeling mentally calmer and more like the version of me that I'd gotten used to over the past twenty-four years.

Since I am learning of new medications and practices that may have made my overall recovery somewhat smoother, I feel slightly guilty. While using such techniques and drugs may have made my healing process somewhat smoother, I also think there's a good chance that I wouldn't have been able to do such investigation on my own. Also, it is commonly agreed upon among medical practitioners that it takes the brain about two years to heal even in the most minor brain injury cases. Since I'd have had to wait two years no matter what, the fact such pieces of information could have really helped me earlier on allow me to shrug it off in a way. I also am able to feel somewhat more accomplished being that I was able to research such information on my own rather than be given medications by doctors.

At this point, though it has been well over one year since my car accident, my days seem to be filled with therapies and doctor's appointments rather than with things I do purely for the sake of my own enjoyment. I am unemployed. I do not drive. I mostly see my mother. My balance is still lacking quite a bit, as is my fine motor coordination. My voice is a disgrace, and my vision is awful even with glasses. I feel as if I'm a useless human being. I often don't interact with friends or even with people my own age. I'm often spoken to as if I'm an ignorant child, and I honestly dislike what is going on with my life. I truly believe that if I didn't have a traumatic brain injury, I would behave in some of the same sorts of ways toward others who act or speak in the way I do either directly to me or near me.

Several Years Later

Just the other day marked the fourth anniversary of my near-death experience. Reflecting on the events of that day in 2015, my mother says it was the worst day of her life. I still live with her and have been doing so for the past four years, since I was discharged from the hospital.

I am still unemployed and unable to drive, and I rarely see or hear from my former coworkers. Though I will say that one of them I have seen and heard from more frequently than the others. There is of course, as previously mentioned, the fact I still have to live with my mother. It makes financial sense as I don't have to pay her any rent. I am also on food stamps, as I have been for a while now, and I don't have to pay out of pocket for food. I have been receiving a small portion of my court case winnings each month for a few months now. Thanks to a different court case, I have finally been approved for Social Security. Since those two sources of income have finally begun to be added to my bank account, I am much more financially stable than I used to be.

Living with my mother continues to be a struggle. Though I was once in desperate need of her help with regard to different aspects of my life, I no longer am. She goes to work for eight hours a day on the weekdays, and I'm alone in the apartment most of the day. I take care of myself in every way. I have no problem taking a shower or shaving my legs when necessary. I get and make my own food. I clean when necessary, and I take care of my little dog.

My dog is my emotional-support animal, and I have a letter from the

Kerry Rois

office of the neurosurgeon that operated on me verifying her as such. She is a mix between a shih-tzu and a Yorkshire terrier, with the body shape of an Italian greyhound. Needless to say, she is a very petite little thing. Since she is so small, I am easily able to manage her. I couldn't say the same for a larger dog. Her name is Lucy, and I love her very much. I've never had a dog before, but I knew I liked them—definitely more so than cats, as I have so far had some bad experiences with them. I tend to like and gravitate toward animals in general.

My Advice

I will only make suggestions based on the hope that you use this information as a basis for your own decisions. As previously stated, what worked for me may not work for you; it is therefore my hope that you keep an open mind with regard to trying out different things that might really help you. Some things may work, and others may not.

What I have learned either are the following tidbits. Working on the physical mouth movements associated with speech is often glazed over by speech pathologists despite the fact such exercises are sometimes of the utmost importance. Seeing a naturopath who can help your bodily health could prove to be a very successful appointment. Neurofeedback as well as biofeedback can be most helpful. There are devices you can purchase for both of these things. However, you can also listen to classical music for free for a certain level of neurofeedback. There are also a variety of inexpensive vitamins and supplements that you can acquire to help with a brain injury. I have learned that neurotransmitters may be negatively affected by such an injury. Deficiencies of neurotransmitters may prove quite detrimental. There are also a number of medicines you can take that increase mental acuity. The pills I am taking right now on a daily basis include one that replenishes a specific neurotransmitter, as well as one that's commonly taken by pregnant women.

Tai chi also seems to help as far as balance and stability are concerned. I tend to wrap up the fingers on my left hand when I practice tai chi because, as I already mentioned, they shake and twitch at times when

I want them to stay still. Therefore, I force them to. I'm not able to accurately say yet if the tai chi is having any profound impact on my physical abilities, though it certainly is a calming activity. And I'm taking steps toward training my fingers to stop shaking at the very least. Apart from that, it's nice because it's slow and thought out; therefore, your body actually has to respond to the movements you are thinking about.

I've also recently become fascinated with aromatherapy and what sorts of specific issues different oils can alleviate. This, combined with the fact I am certainly cognizant enough to do some of the research on my own, has me kind of excited to give aromatherapy a real shot. I have already learned that using frankincense oil works as an antidepressant of sorts. I just recently learned that using rosemary oil can sharpen your mind and also bring your internal body temperature up. That idea is more enticing to me than the other benefit since I not only live in the northern part of the United States but also in a region that's well known for having bad winter weather.

Epilogue

When I was first attempting to decide on a title for the narrative you have just read, I thought of the title of a famous Pink Floyd song. I later changed the title for the sake of avoiding copyright law issues and things of that nature. My desire was to simply convey the idea that for quite a while immediately after the car accident, I was expected to be a pleasant, quiet little girl. I essentially had to not express any feelings other than a general pleasantness, despite the fact I was fairly often spoken to in an insulting way. It was not about me, after all; it was much more about my family members who were "taking care" of me. It was seen as my responsibility to make sure that most people I came into contact with were in and remained in a good mood. While they all felt good about themselves and all they'd done for the sake of a poor, unfortunate handicapped individual like myself, I was screaming and cursing them out on the inside while keeping my mouth closed. As a result, I became all too accustomed to suffering in silence.

There was also the fact therapists and doctors alike came to the conclusion that there wasn't much more that they'd be able to do for me; in their opinions, I was perfectly suited to live an appropriately independent life, and therefore, they might as well call it quits.

At the same time, I feel as though this event that has left me disabled has made my life more interesting. Before this happened, I was generally fairly good at different things that I set my mind to. Not to brag, but I got very good grades in school, I was a very talented singer, I was a good dancer, and I was decent at playing tennis and doing different sports. In essence, various things were just too easy for me. I didn't really need to try too hard, and therefore my life was boring. Granted, I didn't love the way my life was going, possibly because of the wishes and desires of my family members.

Now that this unfortunate event has occurred in my life, I finally have a real chance to do my life over in the sense that I can finally live my life the way I want to. Up to this point, I've mostly done things that considered the wishes of my various family members. I feel as though despite leaving me in a disabled state, the accident has made it possible for me to move on with my life and take it in a different direction.

I changed the title again as I thought about the series of events that have happened in my life in a very general sense. For whatever reason, it seems that I have proven to be prone to bad things. It makes me fear for what or when the next bad situation may be.

Health Consciousness

Most likely influenced by my most recent experience as an in-patient, I hope in the future to only require hospitalization due to externally inflicted trauma. As such, it is my goal to not injure myself internally with the food I put into my body. This obviously includes the foods I am allergic to but also foods that cause damage to not only my body but the human body in general.

Please note, I am not a registered dietician in any sense of the word. While I may have done well in the science courses I took in high school, that was quite a few years ago, and I have lost most of the information I once knew. I do, on the other hand, have a fair amount of personal experience with some of these truths—more experience than the average person would have, at the very least.

Probably as a direct result of the injury I sustained, I have noticed my body and brain react to different foods in peculiar ways. I do not know if this is because my brain is somewhat unique in its construction and how it reacts to various foods or if these things I've noticed as being true for me are also true for all others.

As far as my personal experiences are concerned, I would advise against consuming large amounts of sugar or high-fructose corn syrup. On this subject, I have done a little research and have learned, or possibly relearned, that while sugar may be necessary to provide the body with an adequate amount of energy, it also manipulates hormones in a certain way. Immediately after ingesting a fair amount of sugar, your body will convert it into serotonin. Serotonin is a feel-good hormone that will result in you feeling happy, though only for a short time. Shortly after you feel happy due to your recent excessive sugar intake, you will experience an emotional crash. Rather than your body returning to a normal level of serotonin, your levels of serotonin will actually plummet below their levels before the sugar intake. This extreme drop will cause you to feel even more depressed and pessimistic than you would have if you had never ingested the sugar. The only way to combat this extreme drop in levels of serotonin is to ingest even more sugar in an effort to keep your mood up at an acceptable level. This creates a vicious cycle and often leads to obesity.

While your waist size may very well be of the utmost importance to you, mood is much more important to those you interact with on a daily basis. The way one feels following a crash is not fun in the least. You feel utterly hopeless and as though there is no point at all in continuing. It may lead to suicidal thoughts and/or behaviors.

While small amounts of sugar are fine and even necessary to an extent, avoiding excess sugar in a general sense will cause your body to maintain a certain blood sugar level, without any unnecessary spikes or dips that will negatively affect your mood. You will be much more even in your general temperament.

I feel like I am neither qualified nor learned enough to advise you on which sort of diet to adopt, or which sorts of foods are good and which sorts should be avoided. Everyone's body is unique on the inside as well as on the outside. What I have noticed as being common among various diets is that a smart way to nourish yourself is to avoid foods that are highly processed.

As human beings, we evolved as animals eating plants and other foods that are fresh and natural, not food products that have been heavily altered by manmade machines and chemicals. Therefore, the smart thing to do would be to try to return to a diet that mirrors the early humans' diet as closely as we possibly can. Obviously, we can't all go outside and hunt for an animal we can then eat and gain protein from.

However, paying attention to more natural foods and their nutritional benefits is certainly possible and advisable.

This brings me to my next point: hemp. Hemp is closely related to marijuana, which has been made illegal due to the fact it is often used recreationally. They are not the exact same thing, however. They do come from the same species of plant. The main difference between the two is their individual genetic makeup. The main compounds that are looked at in the case of distinguishing the two from each other have long, super-scientific names. They are commonly referred to THC and CBD. THC is the psychoactive compound, whereas CBD is the beneficial, medical compound. Whereas marijuana tends to contain anywhere from 10 to 30 percent THC, hemp oils and products tend to contain less than 0.3 percent. CBD, on the other hand, generally tends to be fairly heavy in strains of hemp that are meant for medicinal purposes. This is where the benefits are found, so it needs to be abundant in these plants.

Unfortunately, the medicinal benefits of hemp have been ignored recently due to its relation to marijuana. Ancient peoples may very well have known about the benefits of consuming such a plant in different ways. Since human beings have evolved, changed, and become more closely related to what we now consider to be a modern human, hemp has been shunned due to its relation to marijuana. Entirely separate from its close relation to marijuana, hemp has many health benefits; for example, hemp acts as an anti-inflammatory agent internally as well as externally. Apart from this, it has many other health benefits, including, but absolutely not limited to, its ability to influence new cells in an area to replace old, possibly damaged cells. As a medicinal plant, it should be finally appreciated. We, as the modern human race, have wasted too much time ignoring the wonderful medicinal benefits of such a plant. We have been far too concerned with developing Western, pharmaceutically created, chemically laden medicines for various health concerns, most of which have negative side effects. Hemp, on the other hand, is a plant that can be found as is, growing in nature. It has no potential risks or hazards that have been discovered yet.

Among other things, the ingestion of hemp oil can also increase appetite. It's kind of like the stereotypical notion of getting the munchies after getting high on marijuana. And while some may not see the benefit of having the appetite increased, it is necessary for many people. Weight is an interesting issue. While some have to put in effort and must exercise self-control to lose weight, others must do the same to gain weight. While being morbidly obese has its own health issues, being too slender also has health issues. There is a healthy weight that each person should take into account depending on his or her height. While some may exceed their healthy weight, others can't even meet it.

If gaining a certain amount of weight is necessary for you, filling up on highly processed foods that contain unnecessarily high amounts of sugar is certainly not the best route to take. As previously mentioned, when consumed in mass quantities, sugar can and often does have a severe impact on your mental health. Rather than consuming sugar and very sugary foods, there are much healthier foods you can eat to put on some much-needed weight. These types of foods would be things like: avocados, butter, potatoes, chia seeds, salmon, and nuts. I have read of others, but I would prefer that you look into what they are. In essence, it's all about saturation. Whether a fat saturated determines whether it's good for your body. Saturation is less than ideal. In short, unsaturated fats are healthy fats, whereas saturated fats are best avoided.

However, it is also important to note that some of both types of fats are necessary for a successful existence. That being said, since unsaturated is considered to be the healthier fat of the two types, the amount of unsaturated fat consumed on a daily basis should outweigh the amount of saturated fat consumed. One simple step that can be taken to differentiate between the two forms is in their consistencies at room temperature. Saturated fats tend to be in a solid state whereas unsaturated fats tend to be in a liquid state. Though the reasoning behind limiting your saturated fat intake has yet to be officially proven, it is generally believed that it is good for heart health.

To list a few, unsaturated fats can be found within nuts, plant oils, certain types of fish, olives, and avocados. As far as saturated fats are concerned, they can be found in foods like beef, bacon, hot dogs, and breakfast sausages. While consuming foods that contain high levels of saturated fat may be necessary, it is important to monitor and limit your consumption of such.

Keep Your Mouth Shut

Oral health is another important part of overall health. Bad breath shouldn't be on the forefront of your concerns. Tooth decay and degradation should be considered to be of the utmost importance. Then there are things like cavities and gingivitis. Those types of issues are unpleasant to experience and should be avoided if at all possible.

In addition to your maintenance of your teeth and gums, your oral health is also affected by the foods you consume.

About the Author

Since this is an autobiographical account, there is already much you know about the author. At this time, the author is still dealing with certain things. As such, and in addition to those physical realities the author must now deal with, she is not exactly a fan of how her life is "progressing." She puts that word in quotes because it seems as though nothing is actually happening. She is repeatedly told that things are happening, but she is unable to actually see any of the supposed progress. Only time will tell.

www.ingramcontent.com/pod-product-compliance
Lightning Source LLC
Chambersburg PA
CBHW070816220526
45466CB00002B/683